WATER CRYSTAL HEALING

WATER CRYSTAL HEALING

MUSIC AND IMAGES
TO RESTORE
YOUR WELL-BEING

MASARU EMOTO

ATRIA BOOKS
NEW YORK LONDON TORONTO SYDNEY

BEYOND WORDS
HILLSBORO, OREGON

ATRIA BOOKS
A Division of Simon & Schuster, Inc.
1230 Avenue of the Americas
New York, NY 10020

BEYOND WORDS
Beyond Words Publishing, Inc.
20827 N.W. Cornell Road, Suite 500
Hillsboro, Oregon 97124-9808
503-531-8700 / 503-531-8773 fax
www.beyondword.com

Editor: Julie Steigerwaldt
Managing editor: Henry Covi
Copy editor/proofreader: Marvin Moore
Cover, interior design, and composition: Jerry Soga

First Atria Books hardcover edition October 2006

ATRIA BOOKS and colophon are trademarks of Simon & Schuster, Inc.
Beyond Words Publishing is a division of Simon & Schuster, Inc.

For more information about special discounts or bulk purchases, please contact Simon & Schuster
Special Sales at 1-800-506-1949 or business@simonandschuster.com.

The Simon & Schuster Speakers Bureau can bring authors to your live event. For more information or to
book an event, contact the Simon & Schuster Speakers Bureau at 1-866-248-3049 or visit our website at
www.simonspeakers.com.

Printed in China

10 9 8 7 6 5 4

Library of Congress Control Number :2006928982

ISBN-13: 978-1-58270-156-1
ISBN-10:   1-58270-156-3

The corporate mission of Beyond Words Publishing, Inc.:
    Inspire to Integrity

CONTENTS

Disc 2

INTRODUCTION

WHAT IF THERE WERE NO MUSIC?

What if there were no music at all in this world? Can you imagine such a development? No music, no concerts, no records, no karaoke, no CDs. No songs to sing. No music programs on the radio or television. It's difficult to even imagine. Without music, the world would be a boring place. Just thinking about such a world makes me depressed and stressed. Music has become such an integral part of our daily life that we can't live without it.

WHAT IS MUSIC?

In Japanese, we write the word *music* (音楽) as "enjoyment" (楽) of "sounds" (音). What is a sound? A sound invariably occurs wherever there is a vibration, so we could say that a sound is a vibration. (Although humans only hear sounds generated by vibrations between 15 Hz and 20,000 Hz, which are called audible sounds, all vibrations emit sounds.)

What, then, is a vibration? It is energy itself, moving through matter. As waves of energy disturb particles of matter, the particles move, like floating balls on water, going up and down with the wave pattern.

Music, then, is an art form through which we experience vibrations, which is actually energy itself.

THE SYMPHONY INSIDE OUR BODY

Our body is composed of upwards of 100 trillion cells. Surprisingly, the vibration of each cell is slightly different from those of other cells. This means that each cell is generating a slightly different sound. With so many cells in our body, if we could use a high-quality microphone to collect all these sounds, we would hear a great symphony that is constantly being played inside our body.

In my research of vibrations within the body, I have found that the origin of any disease is a disturbance in the frequency within each cell. We could say that the more harmonious and the more beautiful our body's symphony is, the healthier both the body and mind should become. Conversely, when discordant vibrations increase, our body and mind will turn negative and finally a disease will become manifest. For this reason, great doctors of "the good ol' days" were able to find abnormalities in the body of a patient with only the help of a stethoscope.

CONTROLLING A VIBRATION WITH ANOTHER VIBRATION

It is not difficult to restore a distorted sound to its normal state: one can simply measure the wave form of the existing sound, shift its phase ninety degrees (that is, curving "down" where the original is curving "up"), and then play it alongside the original sound. The distorted sound will be cancelled out and no sound will be heard. This technology has already been applied to reduce the noise generated from the engine of a car.

The reason that we need music resides in this principle. The stresses of modern society distort the vibrations of the cells in our body. To assist these distorted vibrations in returning to normal, we choose music—vibrations—with the right rhythm, tempo, tone, and melody to cancel out the distortion. For this reason, music has been our companion since the dawn of human history, in many different forms, and continues to give us healing effects.

"Seeing" Music

Music is meant to be heard. But using the technique of water-crystal photography that I developed years ago, it is possible to "see" music. It works in the following manner: a water crystal is a geometrical design shaped by a vibration, and a sound is a vibration. So if you listen to a piece of music and look at the water crystal created by the vibration of that music simultaneously, you can "see" the vibrational pattern of the music in the crystal and absorb all the vibrations into your body through your eyes and ears.

Music Therapy

Organs and diseases have measurable *hado*, a subtle form of energy that is easily transmittable and present in all things. In English, *hado* translates as "wave motion" or "vibration." In the near future, as research progresses, the hado associated with all organs and diseases will become clear and widely understood. Then, it will become possible to compose music that has vibrations and rhythm corresponding to a certain organ or disease—for example, a piece of music that can cancel the hado of liver cancer.

I imagine that, at some future time, the technology to help people absorb beneficial vibrations will be widely adopted. When that day comes, a water-crystal representation of a piece of music will be placed on the music's packaging, and listeners will be able to enjoy music that is composed of the vibrations that are most beneficial to them. They will select music with the water-crystal image they are most attracted to at that moment and will receive healing effects from "seeing" as well as "hearing" the vibration. While enjoying the music, listeners can not only reduce their diseases but also prevent them.

THIS MUSIC SERIES

Until such a day arrives, I have collected these pieces of classical music, which my research has shown to produce beautiful crystals that suggest healing and well-being.

I'd like to describe the process of this research. We start with the premise that water has an ability to memorize various kinds of information. Because music is a assemblage of vibrations, it can be said that the difference in separate pieces of music is the difference in vibration. Water-crystal photography is a technique that can express the quality of each piece of music with a visual image. The water-crystal photographs in this collection of music were taken after each piece of music contained in the CD was played to distilled water. Here is a brief explanation of how we do this.

1. First, we play each piece of music with a set of speakers to the vial filled with distilled water.
2. We place about 1 ml of the water sample onto each petri dish. We do this with fifty dishes for each water sample.
3. We freeze the water contained in the fifty petri dishes at a temperature of -25 to -30 degrees centigrade.
4. Illuminating the frozen ice on each petri dish under a light microscope, we observe each water crystal and take sequence photographs with a magnification of 100X to 200X. These observations and photographs are made inside a large, cold room at -5 degrees centigrade.

It is the tip of the frozen and expanded ice that produces a water crystal. The ice block that is taken out of the freezer melts gradually because of the temperature difference and the light from the microscope. At this moment, the water crystal at the tip of the ice block shows the expression of water that has resonated with each beautiful piece of music. The technique of water-crystal photography can capture this moment as a picture. We continue to take beautiful images that complement each selection of excellent music we test.

But that is not all. The next step is to measure the hado contained in a water-crystal photograph. Let me explain how I came to measure the hado of photographs. For over fifteen years, I have successfully used a hado-measuring device, called a Magnetic Resonance Analyzer (MRA), to measure the characteristics of hado and adjust the balance of the human energy field. To do a hado measurement on a person, the client places his or her palm on the MRA's measuring plate. I have found that if the client cannot be there in person, a photograph of the client works equally well in getting a hado measurement. Once the hado has been assessed, we can transfer to water a hado that counteracts any negative hado generated with the MRA. This is possible because water can memorize hado, which is vibration. The client then drinks the hado water, which corrects the disturbance in his or her energy balance, activates immunity, and facilitates self-healing power.

Now, let's return to music. When we began to measure the hado of a piece of music, we found it difficult to do so directly, because music has no visible form. However, it became possible to measure the hado of music by measuring the hado of the water-crystal photographs of water that was exposed to the piece of music. By combining the water-crystal photography and hado-measuring techniques, the healing effects of music become very clear and concrete.

My associates and I exposed water to pieces of music, observed the resulting water crystals, and took photographs. At the same time, we measured characteristics of the hado of each photograph with the MRA. Based on the hado data, we could infer the potential healing effects of each piece of music in terms of its hado effect.

In our experiments, we used timeless pieces of music that have resonated well with people for many years. Accompanying the photographs is a brief history of each musical selection and its composer so that you can connect with the emotions and hado of the original composer while you fully enjoy the music.

To use this book and CD set, please read each piece's history and hado qualities first and then enjoy both the lovely music and the corresponding beautiful water-crystal photographs. As you do so, visualize each organ in your body, as described in the commentary, with great appreciation for the organ. The concentration of consciousness created by visualizing each organ and the wonderful, beneficial synthesis between seeing and hearing can impart healing effects unlike any you have previously experienced.

Bedřich Smetana
"The Moldau"

Conducted by Antoni Wit
Polish National Radio Symphony Orchestra

Six connected parts make up Smetana's symphonic poem, *Má Vlast*, or *Images of Homeland*: "Vysehrad," " The Moldau," "Sarka," "From Bohemia's Meadows and Forests," "Tabor," and "Blanik." In them, Bedřich Smetana attempts to convey in musical form the essence of the Czech nation. "The Moldau" is the best known of all the parts that make up *Images of Homeland*.

According to a commentary by the composer, the Moldau River (*Vltaya* in Czech) has its headwaters in two separate sources. These tributaries then join to form a single river that flows through meadows and forests. Along the banks of the Moldau, people can be seen holding festivals or engaging in hunting. When evening arrives, the composer says, the moon comes out and water nymphs start to dance about. The river then passes through one of its rapids, and as the mighty movement of water takes a more leisurely pace, it enters Prague with the ancient castle Vysehrad coming into view.

In the beginning of this piece, the flute and clarinet represent the two sources of the river. Then the first violin and the oboe start to play softly, beautifully conveying the theme of "The Moldau." This is a theme that finds ways to penetrate your soul. As the Moldau is about to flow into Prague, when the old Vysehrad Castle can be seen, the theme of the first part is played again on the woodwinds. The Moldau then carries out its majestic yet calm entry into Prague, the Czech capital.

Smetana started work on the six parts of this symphonic poem, *Images of Homeland*, in 1874, the year that he lost his hearing. The composition reached its final form in 1879, and the premiere performance of all six parts was held in Prague on November 5, 1882. The annual music festival Prague Spring starts on the date of Bedřich Smetana's death, May 12, and without exception, the Czech Philharmonic performs all six parts of *Images of Homeland*, Smetana's legacy to his beloved home.

Track 1

"The Moldau"

Bedřich Smetana

"The Moldau" contains hado that can cancel the hado of irritability. At the same time, it resonates with and thus enhances the hado of the lymphatic system. Based on these results, my research suggests that by listening to this piece of music, you can alleviate irritability and vitalize lymphatic tissues in the body. When you feel irritable, your body is giving you a sign that you need calmness and peace of mind. By treating yourself more tenderly, your mind will become clearer.

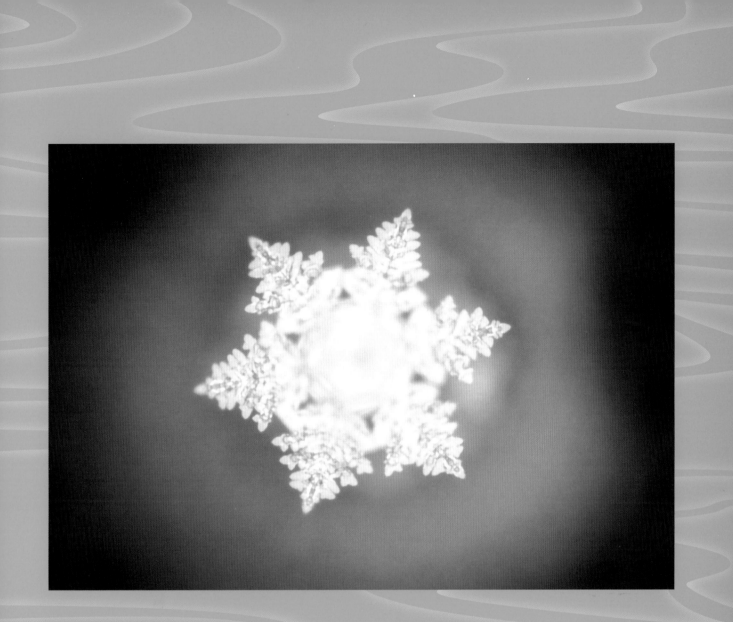

Emotional hado: calmness, peace of mind, relief from irritability

Physical hado: lymph (improved lymphatic flow)

Johann Strauss II
"The Blue Danube"

Conducted by Ondrej Lenàrd
Strauss Festival Orchestra

Johann Strauss I, known as the "father of Viennese waltzes," was the father of Johann Strauss II, who composed an immense number of Viennese waltzes and polkas. The popularity and fame of Johann Strauss II have spread throughout the world. "The Blue Danube," which Strauss wrote at age forty-one, and "Tales from the Vienna Woods" are two of the most popular waltzes ever written.

In 1866, when Strauss wrote "The Blue Danube," continental Europe was dominated by two major powers: France under the rule of Napoleon III and the Austrian Empire under the Hapsburg Emperor Franz Josef I. However, in 1866 Austria found itself engaged in a war against Prussia, which was then in the process of creating a new German Federation. When Italy attacked Austria from its rear, Austria suffered a major disadvantage and went down to defeat.

In July 1866, the disastrous battle of Königgrätz took place between the Austrian and Prussian armies, with the Austrian army losing approximately 100,000 of its 250,000 soldiers. Vienna was filled with masses of wounded troops, and the victorious Prussian flag could be seen flying in Vienna. Finally, the Prussians departed and Austria found itself like an orphan country and had to relinquish control of the Italian city of Venice.

The Austrians could still see the former glory of Vienna and folded their combination of sadness and stress into a dark mood as they attempted to welcome the spring of 1867. But Vienna, surrounded as it is by the glories of nature, witnessed a lively rebirth of happiness. Strauss turned his eye to the boundless, unchanging nature around him. Putting aside the uproar and confusion of the world, he turned to the fast-flowing Danube to capture the soul of Vienna.

Fair lady, enduring the world's pains
You who bask in dignity and youthfulness are
The very foundation of our soul that
Is to be found on the shores of the Danube,

On the shores of the beautiful blue Danube
Oh, nightingale, we hear your song
On the shores of the Danube moving along,
The beautiful blue Danube
　　　　　—from a poem by Karl Isidor Beck

Strauss employed this poem's final line to name his waltz. The composition originally included a song text by Joseph Weyl, and Strauss then adapted it into a purely orchestral version. The composer himself conducted the well-received premiere performance of the orchestral waltz at the 1867 Paris Exposition. After that success, Austria became the second home of this well-loved waltz, and "The Blue Danube" has become a point of special pride for the Austrian people.

TRACK 2

"THE BLUE DANUBE"

JOHANN STRAUSS II

While you are listening to "The Blue Danube," your central nervous system, which may have withered due to habitual constraint, will be revitalized. Your heart will open and your body will feel relaxed. If you have acquired a habit of living in the fast track with a high activity level, your cells are calling out, "You don't need to suffer any longer. Live freely."

Emotional hado: revitalization after fatigue and stress

Physical hado: central nervous system

9

Iosif Ivanovici

"Waves of the Danube"

Conducted by Jerome Cohen
Slovak Radio Symphony Orchestra

Composer Iosif Ivanovici in 1878 was the head conductor of the martial music section of the newly formed country of Romania. He died in Bucharest on September 16, 1902, but little else is known about him.

The score of "Waves of the Danube" is thought to have been written in 1880. During this period, Romania was under the sway of French culture; by extension, Romanian waltz composers were probably influenced by Emil Waldteufel, the "king of the French waltz." What makes Ivanovici's waltzes stand out is his unique ability to combine a major key and a minor key in the same piece.

From olden times, waltzes in a minor key were especially popular as a means to elicit a mood of sadness; such a mood is doubtlessly brought on when one views the last rays of a setting sun reflected off the waters of a fast-darkening Danube. In the United States, Ivanovici's "Waves of the Danube" achieved great popularity after lyrics were written for the melody and the piece became known as the "Anniversary Song."

TRACK 3

"WAVES OF THE DANUBE"

IOSIF IVANOVICI

Listening to "Waves of the Danube" will help you express your emotions honestly and without any suppression. When your heart is open, you help to clear any stagnation in your blood flow and make your cells more vibrant and active than ever before. Besides suppression, other kinds of emotional hados that cause the flow of water (blood) inside your body to stagnate include stubbornness, greed, and excess fear.

Emotional hado: open-heartedness and relief from suppression

Physical hado: promotion of blood circulation

CLAUDE DEBUSSY

LA MER

Conducted by Alexander Rahbari
Belgian Radio and Television Philharmonic Orchestra

TRACK 4

I. "DE L'AUBE Á MIDI SUR LA MER"
(FROM DAWN TO NOON ON THE SEA)

TRACK 5

II. "JEUX DE VAGUES"
(THE WAVES AT PLAY)

TRACK 6

III. "DIALOGUE DU VENT ET
DE LA MER"
(A CONVERSATION BETWEEN THE WIND AND THE SEA)

At the end of the nineteenth century and into the beginning of the twentieth, France was the birthplace and center of Impressionism in painting and music. Claude Debussy is an excellent representative of the Impressionists; his symphonic poem *La Mer*, released in 1905, serves as a fine example of his Impressionist accomplishments.

Looking at Debussy's private life in those years, we see that when he ended a ten-year relationship with Gabrielle Dupont, she was so distraught by Debussy's

14

betrayal that she attempted suicide. In 1899 Debussy married Rosalie Texier, and in 1904 he left Rosalie for Emma, the wife of a banker. The following year, Emma gave birth to the composer's child and Debussy divorced Rosalie and married Emma. *La Mer* was a product of this stormy period in Debussy's life.

La Mer, subtitled "A Symphonic Sketch," is presented on such a scale that it became his grandest orchestral composition. Immediately after completing the opus, Debussy wrote this about it: "This piece of music is nonmaterial and therefore all four legs of it stand firmly on the ground (well, sometimes just on three legs). It cannot be treated as a firmly constructed symphonic poem. No, it holds a special purpose. If its substance were to enter a strict traditional form, I think it becomes nothing. Music is made up of colors and rhythms."

In 1911, Debussy asked in a conversation, "And who might know the secrets of the structure of music? The sounds of the ocean. The majestic curve that separates the ocean from the sky. The wind that whirls up through the leaves. The chirping of birds. All of these phenomena produce a diversity of impressions in us humans. We do not appear all of a sudden to start to form thoughts about these particular things; rather, they become background music that we hear. For me, they bring about a personal kind of musical harmony that I keep to myself."

La Mer is made up of three pieces, each with its own title: "De l'aube a midi sur la mer," "Jeux de vagues," "Dialogue du vent et de la mer." The passage of time along with the changes in the richness of hues available to represent the ocean as a part of Mother Nature come together in *La Mer*, and Debussy's deep yearning to embrace the ocean produced a work of beauty.

La Mer premiered on October 15, 1905, in Paris.

TRACK 4

I. "DE L'AUBE Á MIDI SUR LA MER"

(FROM DAWN TO NOON ON THE SEA)
FROM *LA MER*

CLAUDE DEBUSSY

While you are listening to "De l'aube à midi sur la mer" with a silent mind, you will discover how to solve problems in your relationships. When you become aware of your connections and bonds with other people, you activate emotional circuits in your mind and heart. The vibrational function in your whole brain will be activated as well.

Emotional hado: creation of strong bonds with others and relief
from stress in relationships
Physical hado: improved brain function

 17

TRACK 5

II. "JEUX DE VAGUES"
(THE WAVES AT PLAY)
FROM *LA MER*

CLAUDE DEBUSSY

Try visualizing yourself playing in the ocean waves while listening to "Jeux de vagues." Touched by the greatness of nature, your fixed ideas will be released and you will be able to think more freely. When you have unrestricted thoughts, the hado of your blood will improve and your blood will flow more freely and smoothly.

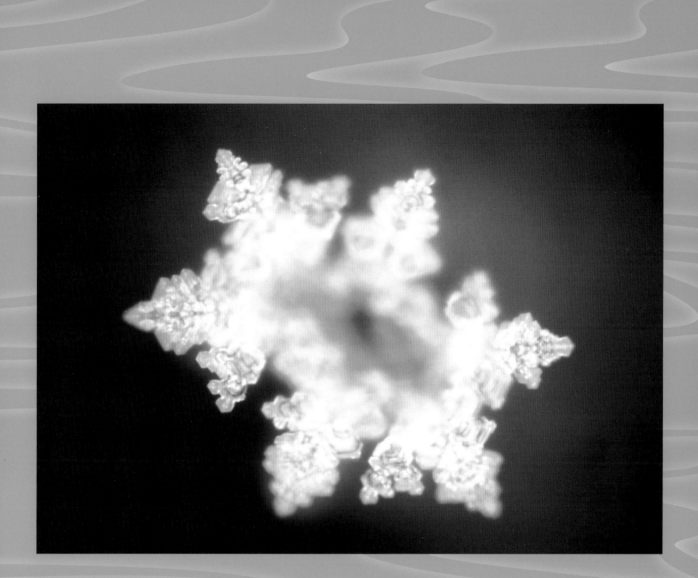

Emotional hado: free thinking and release from obsession

Physical hado: improved blood flow and reinforced platelets

 19

TRACK 6

III. "DIALOGUE DU VENT ET DE LA MER"

(A CONVERSATION BETWEEN THE WIND AND THE SEA) FROM *LA MER*

CLAUDE DEBUSSY

If you listen to "Dialogue du vent et de la mer" repeatedly, the intuitive guidance you receive will help resolve any depression you've experienced because of attachment to past events. By trusting your intuition and progressing forward step by step, your attachment to the past will dissolve and the vibration of the lymph will be activated in your body. Immune system function will improve. You will feel lighter!

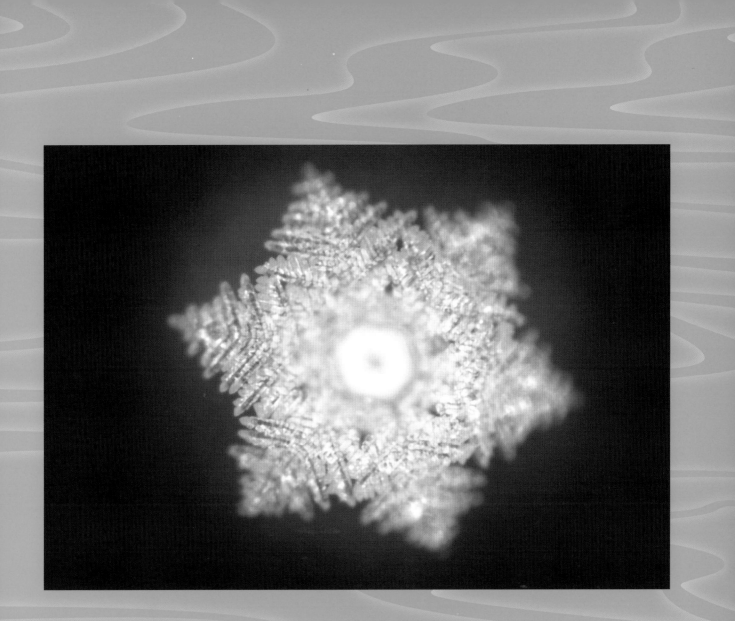

Emotional hado: relief from depression and motivation to move forward

Physical hado: lymph (improved immunity)

 21

Richard Wagner
Prelude to Act 1
from the opera Lohengrin

Conducted by Michael Halàsz
Slovak Philharmonic Orchestra

Among the operatic works of composer Richard Wagner, the one that dazzles in its romantic brilliance is *Lohengrin*. First published in 1848, *Lohengrin*, like *Tannhäuser*, another of Wagner's representative works, depicts a legend from the Middle Ages. In *Lohengrin*, the Germany of the tenth century becomes Wagner's stage. Elsa, the daughter of a lord of a manor, is beside herself in grief after having been accused of a crime that she did not commit. Wagner depicts the Swan Knight, Lohengrin, coming to Elsa's rescue and musically presents the tragic tale of love that ensues between them.

Lohengrin has sworn to protect the Holy Grail. The Swan Knight is capable of wondrous miracles acquired from the powers of the Holy Grail through the blood of Christ and through the powers granted to him through his lineage. This newly arrived Swan Knight gets Elsa to promise that she will never question his origins or lineage, but Elsa later breaks the promise and the knight departs by swan-pulled boat.

In the Prelude to Act I, the violins present the motif of the Grail. Gradually the act reaches a climax, and in a way that defies description, a remarkable beauty is conveyed. Wagner was able to represent the miracle of the power of the Grail through this musical masterpiece.

King Ludwig II of Bavaria first saw this opera when he was fifteen and became so enchanted with its magical powers that he became a patron of Wagner. Wagner then became deeply involved with the construction of Ludwig's castle, the famous Swan Castle, Neuschwanstein, cradled in the mountains looking down on Alpsee Lake. Within the walls of that castle, the nature and sentiment of Wagner's *Tannhäuser* and *Lohengrin* are famous. You can especially understand the extent to which King Ludwig loved Wagner's music by appreciating the meticulous craftsmanship involving the swan motif in Neuschwanstein Castle.

TRACK 7

PRELUDE TO ACT 1
FROM THE OPERA
LOHENGRIN

RICHARD WAGNER

Those who have forgotten the magnificence of themselves often resonate with the hado of self-pity. Respect yourself more. Let us remind ourselves that we are precious beings. Your biggest fan should be yourself. By listening to *Lohengrin*, you will restore self-love. By loving yourself, joint pain will also be alleviated.

Emotional hado: self-love

Physical hado: relief from joint pain

Pyotr Ilyich Tchaikovsky
Scene from Act 2
from the ballet Swan Lake

Conducted by Ondrej Lenárd
Slovak Radio Symphony Orchestra

If someone asked you to name a good example of a ballet, *Swan Lake*, composed by Pyotr Tchaikovsky in 1875 and 1876, would likely come to mind. The foundation for this work is a reverie formed around a German folktale from the Middle Ages.

The story goes like this: A sorcerer changes Princess Odette into a swan. One of Odette's ladies-in-waiting is allowed to change Odette back into a normal person only during the hours after midnight. Prince Siegfried, who has come to the lake to hunt swans, by chance meets Odette in her human form and the two decide to pledge their love to each other.

Meanwhile, a masquerade ball has been scheduled for Siegfried to help him choose a bride. Young ladies hoping to be selected for marriage by the prince have gathered from near and far. The deceptive sorcerer, wanting to terminate the romance between Siegfried and Odette, arranges for the sorcerer's daughter to come to the ball disguised as Odette. When Siegfried discovers the sorcerer's deception, the distressed prince returns immediately to Swan Lake to apologize to the real Odette. However, Odette cannot break the spell that has forced her to take the shape of a swan. Siegfried and Odette are consumed by deep despair and attempt to drown themselves in the waters of the lake. Amazingly, the couple's love is so strong that the sorcerer's spell is broken. Joyfully, Siegfried and Odette are able to marry. (Depending on the producer, the ballet is sometimes performed with a tragic ending.)

In this scene from the second act, the oboe emits a plaintive yet enchanting melody, and this theme is heard throughout the remainder of the piece. The "Swan Lake theme" has entered the realm of popular music. Tchaikovsky takes the moonlight at midnight reflected on Swan Lake and poetically transforms the spectacle of swans grieving over their tragic fate into the music we know so well.

TRACK 1

SCENE FROM ACT 2

FROM THE BALLET
Swan Lake

PYOTR ILYICH TCHAIKOVSKY

Let us behave freely, just as the swan dances on the lake, by tearing

down the walls of fixed ideas that we've built inside ourselves. While

listening to *Swan Lake*, move your body a little more freely. Together

with your body's joints, your thoughts will become more flexible.

Emotional hado: relief from rigidity and obsession

Physical hado: smoother joint movement

JEAN SIBELIUS
"THE SWAN OF TUONELA"

Conducted by Kenneth Schermerhorn
Slovak Radio Symphony Orchestra

Finland's best-known composer, Jean Sibelius, expresses his deep love for the beauty of his native land, blessed as it is with an abundance of forests and lakes. Sibelius manifests his passion for Finland by creating musical compositions that are intertwined with the folklore and history of his motherland. This composition, "The Swan of Tuonela," is one such piece and is the second of four folkloric tone poems in the *Lemminkainen* Suite. Sibelius originally conceived of an opera built around a traditional Finnish epic, *The Kalevala*, and he created "The Swan of Tuonela" as the opera's prelude. However, he was unable to complete the opera and decided to take "The Swan of Tuonela" and breathe life into it, then join it together with three other symphonic poems, and bring the whole composition together in the form of the *Lemminkainen* Suite.

To explain the story of Kalevala simply, Tuonela is the hellish Hades of Kalevala. The Tuonela River unites Kalevala, the mythical land of old, with the Finland of today. Here we find a divine swan singing its plaintive song. A brave warrior, Lemminkainen, accepts an obligation to take the daughter of Pohjola as his bride. Lemminkainen aims to shoot a swan but dies in his failure. Miraculously, through his mother's love, he is brought back from the dead and is able to return to his hometown.

"The Swan of Tuonela" takes inspiration from the shape of the swan. An English horn is the first instrument to be heard, and a special feature of this work is the long breath representing sorrow which then joins with a harp followed by the other stringed instruments creating a mood of dreamy reverie.

TRACK 2

"THE SWAN OF TUONELA"

JEAN SIBELIUS

"The Swan of Tuonela" will make you feel that a promising future awaits you. Whatever pain you may feel, don't give up. Continue to feel hopeful for the future. While listening to this music, visualize your life becoming brighter and brighter. By doing so, the hado in your central nervous system will become balanced and your body relaxed. You will become full of vigor.

Emotional hado: relief from worry and anxiety

Physical hado: acetylcholine (activated neurotransmitter)

CLAUDE DEBUSSY
PRELUDE TO THE AFTERNOON OF A FAUN

Conducted by Alexander Rahbari, Belgian Radio and Television
Philharmonic Orchestra

The name of Claude Debussy can safely be placed at the top of the list of most-loved French Impressionist composers. His *Prelude to the Afternoon of a Faun*, created between 1892 and 1894, was an epoch-making development in the history of modern music. At the first public performance of this work, the audience greeted the never-before-heard sounds with surprise and then, amid calls for an encore, broke into thunderous applause.

Not long before setting to work on *Faun*, Debussy was in communication with the Impressionist poet Stéphane Mallarmé, whose poems enchanted the composer. In 1884, inspired by one of Mallarmé's poetic creations, Debussy composed "Apparition." This was followed by Debussy planning to compose *Afternoon of a Faun* in three parts. As it turned out, Debussy wrote the Prelude. According to Debussy, "This Prelude is in no small measure a free illustration associated with Mallarmé's exquisite poem."

Debussy outlines the contents of Mallarmé's poem as follows: "On a warm, summer afternoon, in a forest, I opened my eyes and beheld the God of Animalism represented by the Faun (an animal god) playing a reed pipe. In the middle, while in a trance, the Faun comes upon nymphs bathing in flowing waters. The Faun thinks of grabbing the nymphs and they disappear. Presently he sees an apparition of Venus and soon the Faun once again sinks back into a deep sleep."

A solo flute plays an inspirational melody at first, languidly. The Prelude then goes on to convey the Faun's carnal thoughts through music.

TRACK 3

PRELUDE TO THE AFTERNOON OF A FAUN

CLAUDE DEBUSSY

Prelude to the Afternoon of a Faun will facilitate the cleansing of your mind, physical body, and fluids of the body. As the water within your body becomes clean, the stress built up around your back will be cleared and your back pain alleviated. Your body will feel light.

Emotional hado: cleaner heart and body and relief from
environmental stress
Physical hado: lumbar vertebrae (improved back pain)

Ottorino Respighi
THE FOUNTAINS OF ROME
Conducted by Enrique Bàtiz
Royal Philharmonic Orchestra

TRACK 4
I. "THE VALLE GIULIA FOUNTAIN AT DAYBREAK"

TRACK 5
II. "THE TRITON FOUNTAIN IN THE MORNING"

TRACK 6
III. "THE TREVI FOUNTAIN AT NOON"

TRACK 7
IV. "THE VILLA MEDICI FOUNTAIN AT SUNSET"

The principal representative compositions of Italian composer Ottorino Respighi are his three Roman symphonic poems/suites. These three suites are *The Fountains of Rome*, *The Pines of Rome*, and *Roman Festivals*. These natural objects and their history, some of which are associated with famous family names, are works that received their inspiration from those family names.

Respighi was also deeply moved by the lessons he received from his teacher, Nikolai Rimsky-Korsakov, and by the splendor of the orchestration of Impressionism as seen in Debussy's law of harmony. Respighi wanted to model his work on Debussy's style that had resulted in high-quality musical masterpieces that were well received by audiences in many different places.

The Fountains of Rome was the first of the three suites and was released to the world in 1916. The composer himself jotted the following comments in the score.

"The Valle Giulia Fountain at Daybreak": "In the middle of the enveloping fog, a herd of cows is moving about creating a calm, pastoral scene."

"The Triton Fountain in the Morning": "The sea-god Triton's presence and the water nymphs' jubilation reach a fullness depicted by their fun and games."

"The Trevi Fountain at Noon": "The sea-god's horse-drawn chariot brings along a parade of mermaids with Triton in its wake, as seen in the surface of the Trevi Fountain's waters."

"The Villa Medici Fountain at Sunset": "A moment's respite at sunset. In the air one can hear the ringing of bells, the chirping of birds, and the trees making a commotion as the wind blows through them."

TRACK 4

I. "THE VALLE GIULIA FOUNTAIN AT DAYBREAK"

FROM *THE FOUNTAINS OF ROME*

OTTORINO RESPIGHI

If you are worried about a relationship, listen to "The Valle Giulia Fountain at Daybreak." Breathe according to the rhythm of this music. This music has a rhythm that will facilitate smoother and more effective communication between you and your family members, friends, and co-workers. When you incorporate the hado of this music into your body, everything will flow smoothly and successfully, including the blood in your circulatory system.

Emotional hado: strengthened bonds with others and relief from stress in relationships

Physical hado: improved blood circulation

TRACK 5

II. "THE TRITON FOUNTAIN IN THE MORNING"
FROM *THE FOUNTAINS OF ROME*
OTTORINO RESPIGHI

Listening to "The Triton Fountain in the Morning" will help heal

the sorrow that resides deep inside your heart and will make you feel

bright and cheerful, like the sun. As the cloud covering your heart

is cleared, your sight will open wider and you will see everything

brightly and clearly.

Emotional hado: cheerfulness and relief from sorrow

Physical hado: improved eyesight, especially farsightedness

III. "THE TREVI FOUNTAIN AT NOON"
FROM *THE FOUNTAINS OF ROME*
OTTORINO RESPIGHI

"How unlucky I am!" When you get depressed about your lot in life, listen to "The Trevi Fountain at Noon" and ignite a light in your heart. Say to yourself, "I can surely become happy." You will have such thoughts more and more. Imagining a bright future will raise the hado of your immunity dramatically.

Emotional hado: good fortune and optimism

Physical hado: improved immunity (antivirus function)

TRACK 7

IV. "THE VILLA MEDICI FOUNTAIN AT SUNSET"
FROM THE FOUNTAINS OF ROME
OTTORINO RESPIGHI

By listening to "The Villa Medici Fountain at Sunset," you will make the useless feelings associated with hardheadedness and pride promptly disappear. These feelings are blocks in your heart that cause stagnation in the flow of "heart energy," or love. When the blocks are removed, you feel relieved and the energy will flow freely. The hado of your thymus, which is an organ that produces lymphocytes, will be activated, and you will surely feel energized.

Emotional hado: relief from stubbornness and pride

Physical hado: thymus (improved immunity)

Richard Wagner
Siegfried Idyll

Conducted by Johannes Wildner
Polish National Radio Symphony Orchestra

As a composer of operas, Richard Wagner has bequeathed to future generations an immense number of works. When it comes to his non-operatic musical compositions, one of his best-known works is *Siegfried Idyll*. At the time of composing *Siegfried Idyll*, Wagner was married to his second wife, Cosima, herself the daughter of composer and pianist Franz Liszt. The Wagner family resided in a manor that he had had built at Tribschen in the natural splendor of the shores of Lake Lucerne. Cosima gave birth to a son in 1869. Wagner was so pleased at this turn of events that he composed a symphonic birthday present, which later came to be known as *Siegfried Idyll*, that he presented to his wife on the anniversary of her birthday, December 25.

On that special morning, Wagner, as conductor, arranged to have a fifteen-member orchestra gather quietly at the family manor. Wagner kept the event secret from Cosima until the musicians started to play. Total surprise was achieved and Cosima was reportedly in bliss to hear her husband's musical creation. At the time that Wagner wrote *Siegfried Idyll*, he was working on a musical drama, *The Valkyrie*, and it is said that he used the motif from the latter in constructing Cosima's birthday gift. The name of "Siegfried," the child who caused Wagner's heart to burst with joy, was attached to this piece of music, which has as its center the waters of Lake Lucerne. As we listen to Wagner evoking the warm feelings of his well-founded happiness, we can sense the abundant blessings of nature to be enjoyed at Tribschen.

TRACK 8

SIEGFRIED IDYLL

RICHARD WAGNER

Siegfried Idyll will raise your inner passion and willingness. As you

focus on your dream and ideal, you become powerful and attractive.

Hot blood will flow in your arteries. This energy of passion will make

your life more enjoyable.

Emotional hado: passion and willingness

Physical hado: arteries (improved blood circulation)

AFTERWORD

By combining water-crystal photographs and hado measurements, I intend to focus and raise the healing effects for readers. This is the first attempt in the world to do this. I sincerely hope you have enjoyed this collection of music and water-crystal photographs.

I believe that hado (vibrational) medicine will become a primary form of medicine in the world someday, and I hope that day comes soon so many people can be helped. Therefore I decided to make my ideas materialize and publish this product.

The water-crystal photographs in this book were taken by Takayuki Oshide at the IHM General Institute. I dedicate my gratitude and respect to his sincere technique of photography.